LIGHT

Created by Gallimard Jeunesse,
Jean-Pierre Verdet,
and Gilbert Houbre
Illustrated by Gilbert Houbre

A FIRST DISCOVERY BOOK

Cartwheel
·B·O·O·K·S·®
SCHOLASTIC INC.
New York Toronto London Auckland Sydney

The sun gives the earth heat and light.
Without light we could not see.

Light is energy that flows
in waves. When light hits objects, we see
the objects in color.
When sunlight passes through rain,
we may see a rainbow.

Green plants need sunlight to grow and make food.

Here's an experiment to try.
Plant some seeds and wait until
they sprout. Put a flower pot over half
of the sprouts. Those will grow in the dark.

What will happen to the tiny plants?

The plants that grew in the dark are white and yellow. The plants that grew in sunlight are green with chlorophyll.

Chlorophyll allows plants to make food from sunlight.

Whether
they live
on land or in
the sea,
plants and
animals need
light to live.

The sun shines on
the surface of the ocean.

Tiny vegetable plankton
make food from the sun.

Small animal plankton
eat the tiny plants.

Fish, such as herring,
eat the small animals.

Bigger animals, such as
penguins, may eat the herring.

Even bigger animals, such as
whales, may eat the penguins.

But sunlight does not reach the bottom
of the ocean. How do the animals
that live there get food from light?
They feed on plankton and dead fish
that are deposited in the mud.

Sunlight affects the food
all animals eat.

Shadows are made when light can't pass through
a solid object or shape. When the shape or light moves,
the shadow changes.

Look at the big rabbit shadow
on the right. How do you
think it was made?

With hands!
You can make hand shadow puppets, too.
Shine a bright light on the wall.
Block the light with your hands.

Move your hands in different ways

to create different animal shapes.

Light moves
in a straight line.
It can't go around or pass through
your hands — so your hands
make a shadow.

Fluorescent plants and
animals glow naturally.

Certain luminous substances absorb
light — then give it off slowly in the darkness

They light up!
A chemical reaction
in the firefly's light
organ produces a
heatless light.

The numbers
and hands of this
clock glow in the dark
because they are
coated with a
phosphorescent
substance.

What do fireflies do at night?

Sometimes we see lights and colors in the sky.

A nebula is a cloud of gas
lit up by a very hot star.

A comet's tail is made of gas and dust.
It is lit up by sunlight.

When light passes through mist in the sky,
we can see shapes of light called halos.

Auroras are
nature's finest light shows.
In the far north they
are called the
aurora borealis or the
Northern lights.

We can't see color in the dark.

When we darken a room,
we see shapes blending together
to make silhouettes.

But when we let
the light back in,
we see separate shapes
and colors again.

Before electricity, fire was used for light at night.

People carried lanterns
with flames  inside.

In 1879,
Thomas A. Edison
invented
the electric
incandescent
lightbulb.

This kind of lightbulb is called
a fluorescent lightbulb.
Today electricity lights
our homes and streets.

At sunset,
the sky gets
darker
and
darker
until…

the lights
go on!
The city
sparkles
with light.

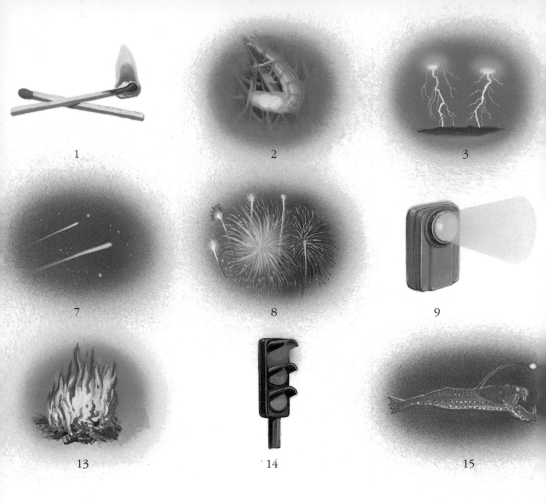

How many of these lights have you seen?
Which ones can you name?
Which lights are made by people?
Which lights are made by nature?

4

5

6

10

11

12

16

17

19

1 matches
2 glowworm
3 lightning
4 halogen lamp
5 sun
6 lighthouse
7 meteors
8 fireworks
9 flashlight
10 flash camera
11 television
12 car headlights
13 campfire
14 traffic light
15 lantern fish at ocean bottom
16 lighter
17 rainbow
18 laser
19 infra-red remote control

18

The sun has set, but its light
bounces off the full moon to light
our nighttime world.

Titles in the series of *First Discovery Books:*

Library of Congress Cataloging-in-Publication Data available.

Originally published in France under the title *Lumiere* by Editions Gallimard.

ISBN 0-590-48327-7

Copyright © 1992 by Editions Gallimard.
This edition English translation by Jennifer Riggs.
This edition American text by Jean Marzollo.
All rights reserved. First published in the U.S.A. in 1994 by Scholastic Inc. by arrangement with Editions Gallimard.

CARTWHEEL BOOKS ® is a registered trademark of Scholastic Inc.

12 11 10 9 8 7 6 5 4 3 2 1   4 5 6 7 8 9/9

Printed in Italy

First Scholastic printing, September 1994